**Dennis Sitsofe Anyomi**

# Power Quality and Industrial Performance

Anchor Compact

Anyomi, Dennis Sitsofe: Power Quality and Industrial Performance. Hamburg, Anchor
Academic Publishing 2015

Buch-ISBN: 978-3-95489-355-3
PDF-eBook-ISBN: 978-3-95489-855-8
Druck/Herstellung: Anchor Academic Publishing, Hamburg, 2015

**Bibliografische Information der Deutschen Nationalbibliothek:**
Die Deutsche Nationalbibliothek verzeichnet diese Publikation in der Deutschen
Nationalbibliografie; detaillierte bibliografische Daten sind im Internet über
http://dnb.d-nb.de abrufbar

**Bibliographical Information of the German National Library:**
The German National Library lists this publication in the German National Bibliography.
Detailed bibliographic data can be found at: http://dnb.d-nb.de

© Anchor Academic Publishing, ein Imprint der Diplomica® Verlag GmbH
http://www.diplom.de, Hamburg 2015
Printed in Germany

# DEDICATION

I dedicate this work to my beloved family. I love you all.

# Table of Contents

# LIST OF TABLES

# LIST OF FIGURES

# PREFACE

I am very excited about writing this book and thank the Almighty God for his constant love, infinite wisdom, guidance and strength to be able to complete it.

Even as a little child I was known as one who questions virtually everything and the phrase "He likes asking questions in class" was one constant remark that appears on my report cards. I believe every system can be made better if the right questions are asked and the solutions to challenges are pursued with the right attitude, knowledge and determination.

Africa has experienced its fair share of power crisis. Quality electricity supply propels industrial growth and boosts economic development, hence power crisis are a dent on industrial development and economic growth at large.

Ghana over the past years has had its own share of electricity crises which in recent years is humorously referred to as "dumsor dumsor" meaning the unreliable nature of electricity supply to households and industries.

I believe my years of experience in the energy sector as an engineer and my postgraduate work in engineering and management puts me in a position where I can understand the issues, reviews current scientific and technical know-how on the subject and share my perspectives on the way forward.

Throughout this book, I intended to share my knowledge, experience, research and thought with cherished readers, starting with basic concepts of electricity, some trends in electricity generation, supply and consumption, the vital elements of quality electricity supply, some general key performance indicators of industries, and how electricity impacts on industrial performance.

This book aims to offer a good knowledge on the subject for all readers hence it does not include technical details, or theoretical and mathematical formulae but has vivid graphs, tables and diagrams explaining and showing trends on issues I hope to carry across.

# INTRODUCTION

Electric current can simply be referred to as the rate at which charges flow across any cross sectional area.

Electricity is the fastest growing form of delivered energy in the world. That notwithstanding, virtually every economy has experienced its own share of energy crisis and the vital implications of poor quality electricity generation and supply on both individuals lives and businesses.

In recent years in Ghana, one word that resonates in virtually every daily conversation from the ordinary citizen on the street, the media houses, through to investors is "power". By power they simply mean quality and reliable electrical energy for homes, schools, work places etc.

Electricity is a vital ingredient for any economy seeking growth in this era of global competitiveness. There indeed exists a strong positive correlation between electricity usage and economic growth and development. However, one key element that most studies fail to note is that, electricity availability is not the sole panacea for industrial performance and growth of any economy.

Quality electricity supply goes beyond mere availability of electricity as most texts fail to clearly distinguish between availability and quality electricity supply. Quality electricity generated and supplied to households and industries does have implications for the energy policy, demand site management and the generation mix of any country.

In this book four (4) general key performance indicators have been identified and discussed. Three (3) general quality elements of electricity have also been discussed. Six (6) vital areas that are impacted by electricity have also been identified in this book.

# PART I

## ELECTRICITY IN CONTEXT

### Brains Behind Electricity

Thomas Edison and George Westinghouse are the brains behind the electrical energy we utilize today in our homes and businesses. This was achieved through a simple electromagnetic power generators and a complicated distribution system.

Benjamin Franklin is known to be the person who demonstrated that lighting is electricity. Generation, transmission and the use of alternating current were pioneered by Nikola Tesla. Charges in the clouds that are being triboelectrically generated and discharged as lighting were harnessed by Benjamin Franklin. Captured electricity was shown in Leyden jars. Whilst Edison was known to have pioneered electricity generation and distribution as well as the light bulbs which made electricity became a necessity in homes, streets and businesses. Maxwell Ampere and others also worked to enhance the understanding of the laws and nature of electricity.

### Electricity Generation

Electricity generation is the production of useful current at some voltage. Alternating current can be produced through a varying magnetic field and the subsequent collection of electrical current from loops of wire.

Electricity generation globally has being increasing, as demand is on the rise. Alternates to energy generation and management are sort to match the rising demand.

World net electricity generation is projected to increase by 93% from 20.2 trillion kilowatt hours in 2010 to 39.0 trillion kilowatt hours in 2040 as shown in the figure below.

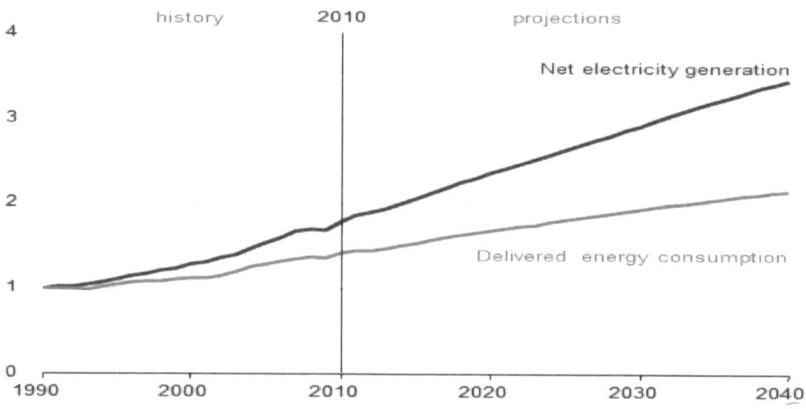

**Figure 1:World Net Electricity Generation in trillion ($10^{12}$) kilowatt hours (Source: EIA, 2013)**

In Ghana the total electricity generated in the year 2002 was 7,273GWhr. A decade after (2012), the total electricity generated was 12,024GWhr and this represented 65.3% increase in the electricity generated over the period.

**Forms of Electricity Generation**

Using fossil fuel for electricity generation is considered the conventional electric power source and the majority of grid electricity is generated through the use of electromagnetic generators by the combustion of fossil fuels. Sources of electricity generation such as harnessing flowing water, light, wind, thermal gradients and any other renewable fuel to produce electricity is therefore referred to as alternative energy.

According to EIA (2009), fossil fuel sources such as Coal, Crude oil and Natural gas together generated more than 65% of the world's net electricity generated.

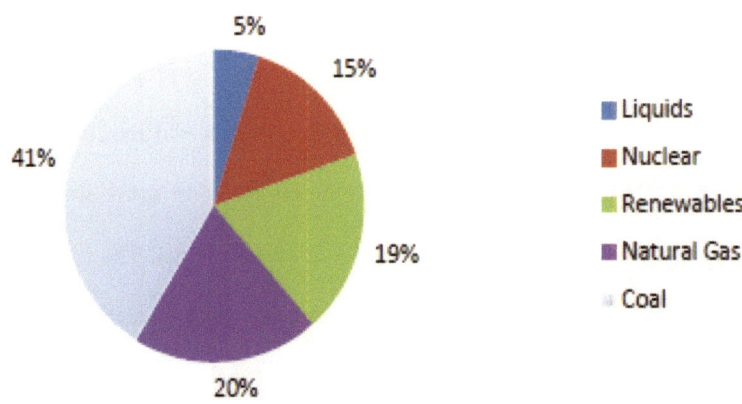

**Figure 2: World net electricity generation by energy source (Source: EIA, 2006)**

As of 2011, about 80% of the world's electricity energy is supplied by fossil fuels. This is largely as a result of the uneven nature of raw material resources use in electricity generation are distribution in the world and this creates energy security challenges. It is therefore accurate to consider all other sources of electricity generation as alternative to fossil fuel.

The conventional electric power sources that use fossil fuel such as crude oil, coal or natural gas poses environmental challenges and hence alternative sources such as renewable energy technologies are beginning to play key roles in the world wide electrical power production. Strategic choices with regards to investment into renewable energy must be considered.

Increases in investment into renewable energy technologies could have dual benefits such as helping meet further energy demands and in minimizing the risk that conventional energy supply poses.

It is refreshing to know that renewable energy is projected to be the fastest growing form of electricity generation in the world by 2040. According to EIA (2013), Renewable sources are projected to grow from 4.18 trillion kilowatt hours in 2010 to 9.60 trillion kilowatt hours. The non-hydro sources of renewable energy are the fastest growing energy generation source among the renewables. The Table below shows some details of the renewable energy projection made by EIA.

Table I: World net projection for renewable electricity generation in billion kilowatts hours (Source: EIA, 2013)

| World | 2010 | 2015 | 2020 | 2025 | 2030 | 2035 | 2040 | Average Annual Percentage change, 2010 - 2040 |
|---|---|---|---|---|---|---|---|---|
| ydroelectric | 3,402 | 3,805 | 4,452 | 4,762 | 5,177 | 5,692 | 6,232 | 2 |
| Vind | 342 | 767 | 1,136 | 1,383 | 1,544 | 1,694 | 1,839 | 5.8 |
| eothermal | 66 | 112 | 133 | 146 | 171 | 195 | 220 | 4.1 |
| olar | 34 | 157 | 240 | 288 | 327 | 394 | 452 | 9.1 |
| ther | 332 | 427 | 549 | 643 | 729 | 800 | 858 | 3.2 |
| otal World | 4,175 | 5,267 | 6,509 | 7,222 | 7,948 | 8,775 | 9,601 | 2.8 |

In Ghana (2013), data shows similar trend in renewable generation sources. Per available data, the dominant form of electricity generation in Ghana is Hydro. The current installed capacity in Ghana is 2,844.50 MW. Out of this, about 55.44% is from hydro generation whilst 0.8% is from Solar. - Figure 5 shows the generation by source in Ghana over the decade whilst Table 2 shows the details on the various generating facilities and their installed capacity in Ghana.

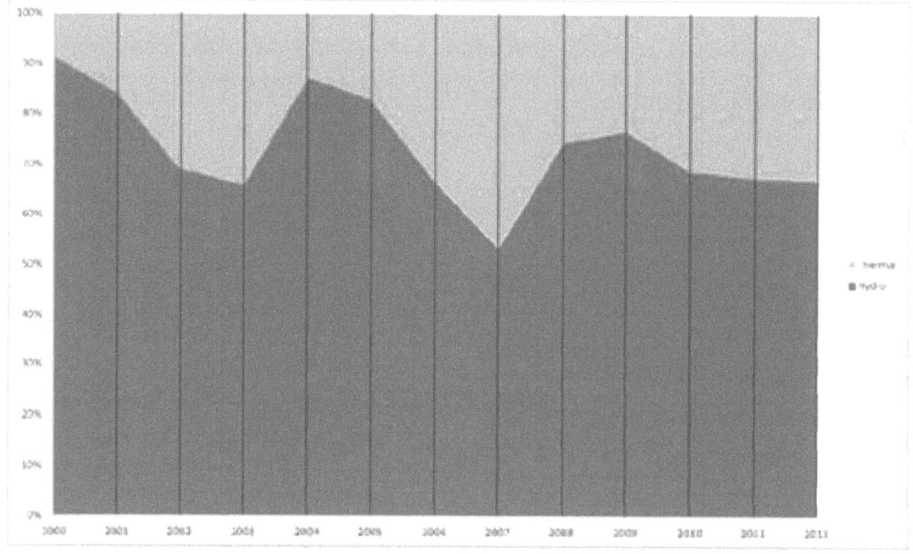

Figure 3: Trend in electricity generation by source in Ghana (Source: Energy Commission, 2013)

Table II: Generation facilities in Ghana and their respective installed capacity

(Source: Volta River Authority, 2014)

| Generation Facility | Installed Capacity (MW) | Installed Capacity (%) |
|---|---|---|
| Akosombo Hydroelectric Power Plant | 1020 | 35.8% |
| Kpong Hydroelectric Power Plant | 160 | 5.6% |
| Takoradi Thermal Power Station (T1) | 330 | 11.6% |
| Takoradi Thermal Power Station (T3) | 132 | 4.6% |
| Takoradi Intl' Company (TICO/T2) | 220 | 7.7% |
| Tema Thermal 1 Power Plant | 126 | 4.4% |
| Tema Thermal 2 Power Plant | 50 | 1.75% |
| Mines Reserve Power Plant | 80 | 2.81% |
| Solar Power Plant | 2.5 | 0.8% |
| Sunon-Asogli Power Plant (SAPP)* | 200 | 7.03% |
| CENIT* | 125 | 4.3% |
| Bui Hydroelectric Power Plant | 399 | 14.02% |
| **TOTAL** | **2844.5** | **100.0%** |

* Independent Power Producer

10

Hydro is a common form of renewable energy. Hydro power produced about 3,288 TWh which forms 16% of global electricity produced in 2008. The overall hydro power technical potential is estimated to be over 16,400 TWh/yr. - Figure 4 below shows the shares by country in hydropower generation in 2008 whilst Figure 5 shows the evolution of global hydropower generation between 1990 to 2008.

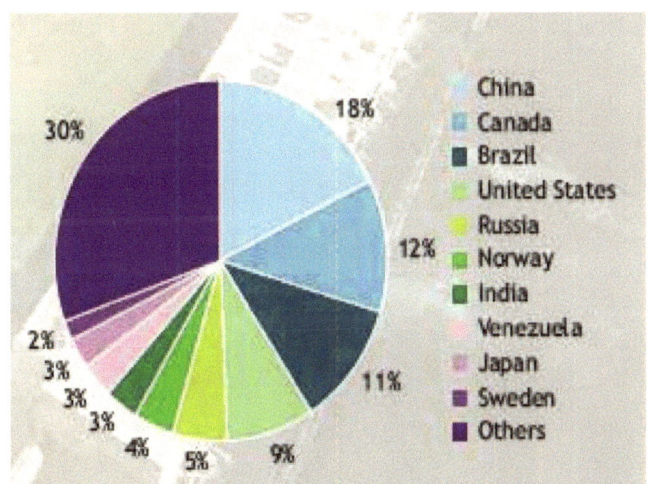

**Figure 4: Share of hydropower generation in 2008 (Source: IEA, 2010)**

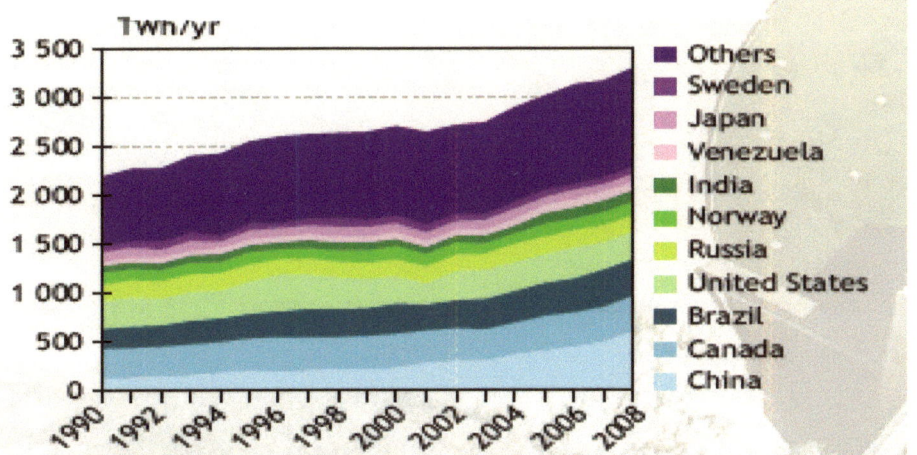

**Figure 5: Evolution of global hydropower generation between 1990 - 2008 (Source: IEA, 2010)**

Coal on the other hand is known to be the dominant fuel used for electricity generation in the world. In 2010 coal fired electricity generation accounted for 40% of the world electricity generated.

Reports however showed a slow increase in the projection of coal fired generation from 8.05 trillion kilowatt in 2010 to 13.80 trillion kilowatt in 2040. Nonetheless coal still remains the largest source of electricity generation through 2040 using EIA 2013 report as shown in Figure 6 below.

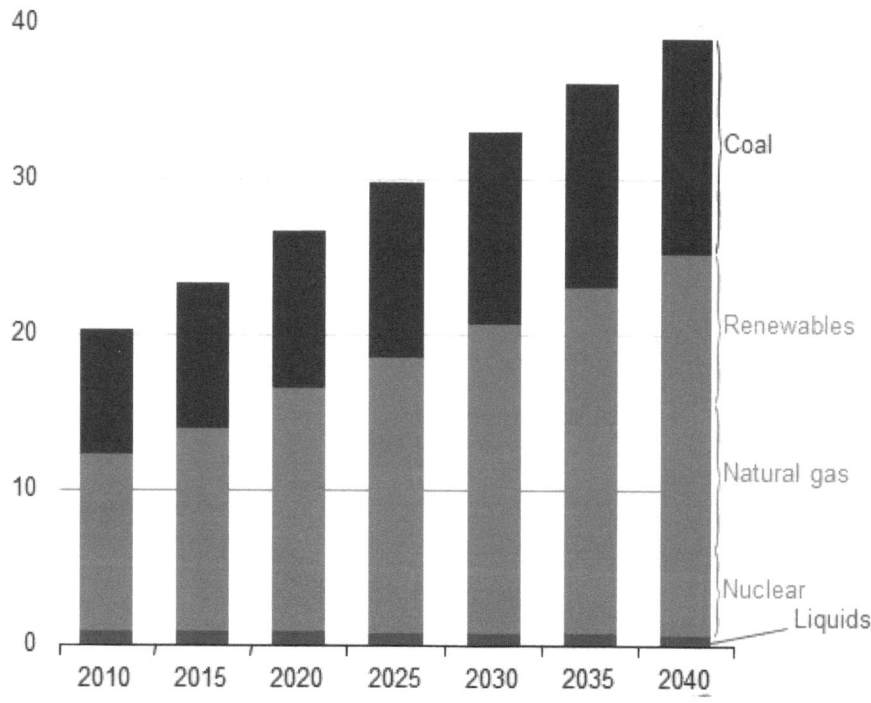

**Figure 6: World net electricity generation by fuel type in trillion ($10^{12}$) kilowatt hours (Source: EIA, 2013)**

Natural gas is noted to account for 22% of the world's energy generation in 2010. Its projection however has improved largely because of the revised expectation of various natural gases.

Nuclear power generation worldwide that produced 2.62 trillion in 2010 is projected to increase to 5.492 trillion kilowatt hours.

Fossil fuel especially coal fired thermal plants appear to be the most dominant source of electricity generation around the globe. This is largely due to the availability and the low cost of coal as a raw material for electricity generation.

Considering the environmental challenges that this conventional energy generation pose, more investment in the renewable sources of electricity generation needed to be considered. This however remains an illusion. On the isolated island in Flores, three (3) sources of generation are present. These are the Hydro, Wind, and Diesel fired plants, illustrated in Figure 7.

Figure 8 below shows the detail installed generation capacity in China as of 2004 and the actual electricity generated from these sources.

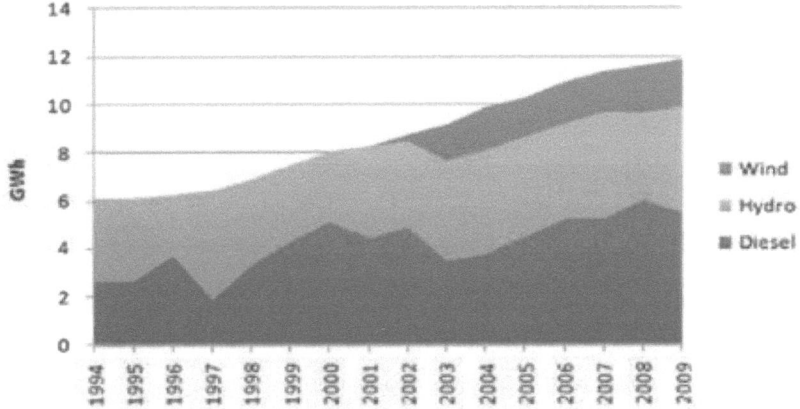

**Figure 7: Electricity Production in Flores by Source (Source: Pina et al, 2012)**

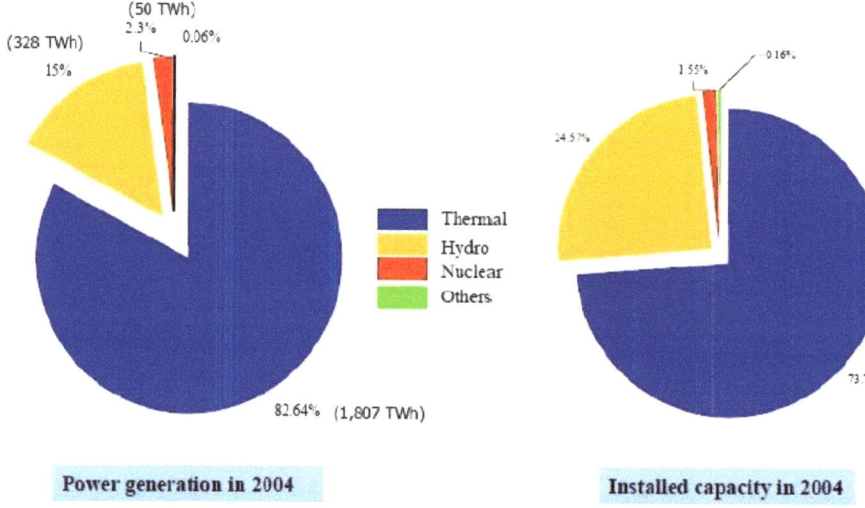

| | |
|---|---|
| Power generation in 2004 | Installed capacity in 2004 |

**Figure 8: China Installed capacity and Actual Electricity Generation in 2004 by source (Source: Jing Li and Yu Xue, 2009)**

In 2013, China was known to have the largest installed generation capacity in the world. It has the largest thermal power capacity, the largest hydropower capacity, and largest wind power capacity and among the countries with the largest solar power capacity. - The total install capacity in 2013 was 1,247GW with coal being about 801 GW, hydropower installed capacity of 280 GW, nuclear of 15.69 GW. The Figure 9 shows the generating capacity by fuel type from 2000 to 2013 for China.

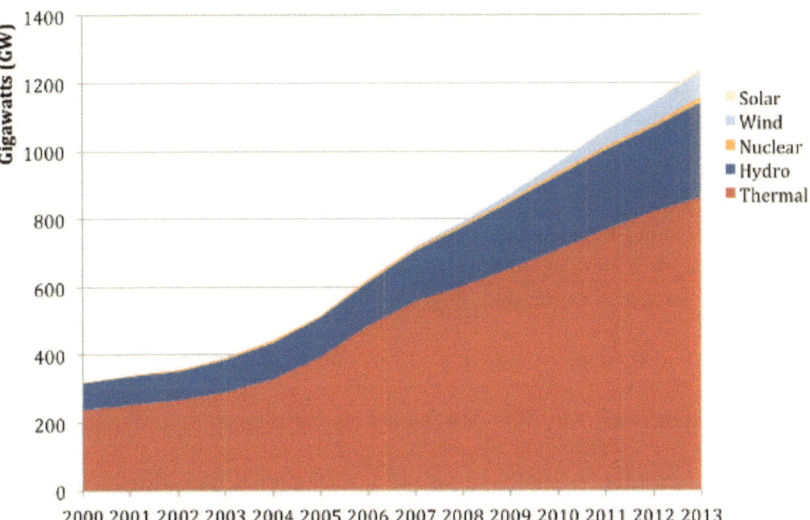

Figure 7: Electricity generating capacity by fuel type from 2010 - 2013 (Source: CEC, NEA)

## Electricity Supply and Consumption

Electricity supplies an increasing share of the world's energy demand.

Increasing energy consumption is one of the concerns that societies are facing in recent times. - Incessant population growth, booming technology, societal growth and increased comfort are some of the reasons.

Globally, energy demand is increasing and the industrial sector uses about one-half the world's total delivered energy.

Electricity consumption has a positive effect on economic growth of any country. One key way this is achieved is by directly improving industrial performance thereby bringing growth in the industrial sector. There are studies that suggest that the adoption of policies to conserve electricity may unwillingly decline economic growth.

In China, two main power grids are responsible for the transmission of power in China whilst and five (5) are responsible for generation.

In Ghana, the Ghana Grid Company Limited is the sole company responsible for transmitting the power generated by the generating companies.

China's energy consumption grew by 7.5% in the same year to 5.3 trillion kilowatt hours KWh). The growth in consumption was generally due to industrial sector which had consumed 3.9 trillion kilowatt hours out of the 5.3 trillion that the entire country had consumed.

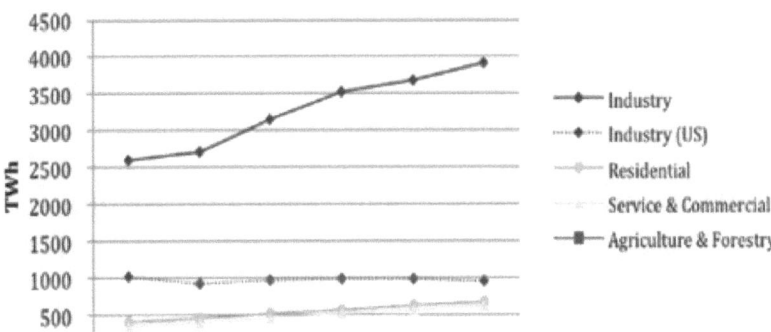

**Figure 8: China Electricity Consumption by Sector from 2009 - 2013 (Source: NEA, 2013, US EIA)**

The state owned electricity generating company of Ghana, Volta River Authority generates and supplies 11,346.02 GWh which represents 88.3% of the entire nation's 7,870.19 GWh in 2013. Figure 11 below shows the detailed electricity supply for Ghana in 2013.

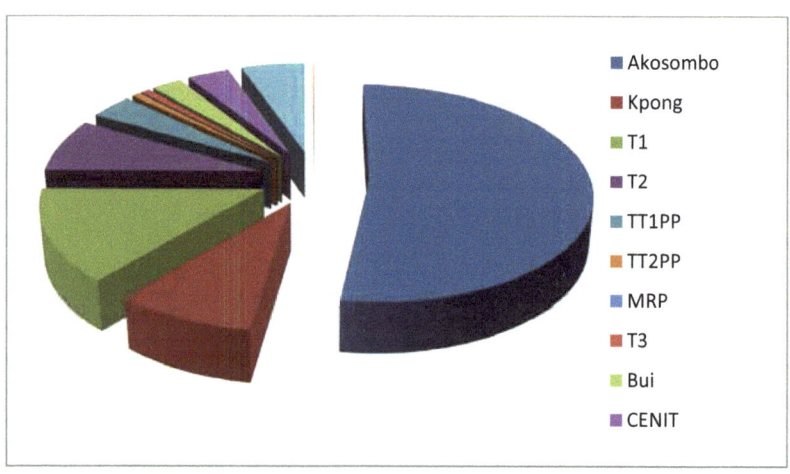

**Figure 9: Annual electricity supply in Ghana (Source: Volta River Authority, 2014)**

**Legends:**
| | | |
|---|---|---|
| Akosombo - | Akosombo Generating Station |
| Kpong – | Kpong Generating Station |
| T1 - | Tarkoradi Thermal Power Station 1 |
| T2 - | Tarkoradi International Company (TICO/T2) |
| T3 - | Tarkoradi Thermal Power Station 3 |
| TT1PP - | Tema Thermal 1 Power Plant |
| TT2PP - | Tema Thermal 2 Power Plant |
| MRP - | Mine Reserve Power Plant |
| CENIT - | CENIT Energy Limited |
| Bui - | Bui Power Authority |

Figure 12 below shows Ghana electricity demand and supply gab. It revealed that there is virtually no electricity reserve margin for Ghana from 2004 to 2008.

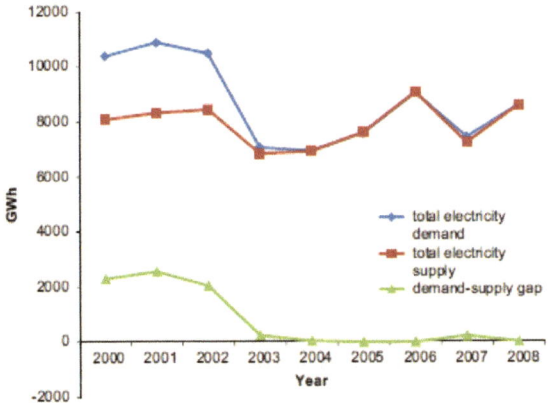

**Figure 10: Demand and Supply Gab in Electricity for Ghana from 2000 to 2008 (Source: Adom et al, 2011)**

Electricity consumption in Flores, an isolated island was largely due to domestic, commerce and service.

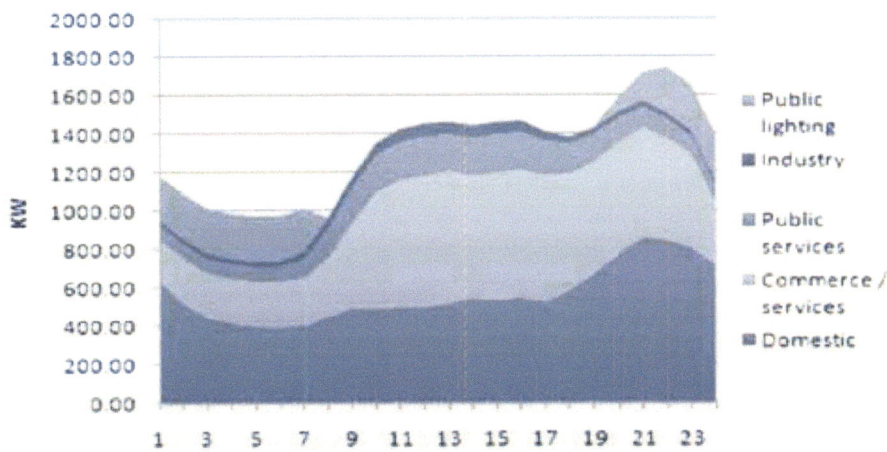

In Ghana electricity demand is driven largely by increasing domestic demand than

**Figure 11: Hourly Consumption of Electricity in Flores (Source: Pina et al, 2012)**

industrial demand. The nation's electricity consumption increase by 35.6% from

2002 to 2012, the load at peak increased within the same period by 88.6%. The industrial consumption only accounted for an increase of 6.4% within the same period.

Also, between 2012 and 2014, electricity demand in Ghana grew averagely at 10% - 15% per annum and therefore a major concern to energy generating sector considering the current Ghanaian installed capacity of 2,844.50 MW vis avis the investment needed to match the growing energy needs of the country. In summation, energy demand will globally continue to increase hence the need to secure more sources of energy and supply but in a more environmentally friendly manner.

# PART II

## QUALITY Electricity

**Quality**

International Standard Organization (ISO) 9000 defined quality in as the extent or degree to which a given set of inherent characteristics fulfills the stated requirement. By this, ISO defines quality as the fitness of purpose. Quality electricity therefore will be the measure of the degree the inherent characteristic that makes up electricity fulfills its requirement.

Lots of individuals and literature have described quality electricity as one that is supplied when there are minimum numbers of interruption during its service duration. Others simply say quality electricity is the availability of electricity as well as the safe and satisfactory operation of all connected devices.

In as much as availability is a vital element in defining quality electricity, in the broader sense, quality electricity supply goes beyond just availability. The issue of voltage quality and other commercial services are also equally vital roles in determining the quality of electricity supply.

The 4th benchmarking report on quality of electricity supply by the Council of European Energy Regulators (CEER) concurred to this by identifying three main factors affecting the quality of electricity. These include the continuity of supply (availability), voltage quality (usefulness of available electricity) and commercial quality.

21

These three main quality elements of electricity supply can be said to be well within ISO 9000 definition of quality in relation to electricity. These three factors therefore can be used to determine if the quality of electricity supply to households and industries are within the acceptable quality measures.

**Continuity of Supply (Availability)**

Continuity of electricity supply deals with the availability of electricity. This means that unavailability of electricity will mean interruption; any situation that makes supply of electricity unavailable to customers. Optimal supply of electricity can vary from region to region (say an urban area and a rural area) and from customer to customer (say a domestic customer and an industry).

Availability as a quality measure is very vital for both the customer and the supplier. In Ghana, for instance an estimated 2 to 6% of Ghana's GDP is lost annually due to insufficient wholesale power supply.

Interruption as stated earlier is the unavailability of electricity. This can be seen, measured and analyzed in two ways. The first is when the magnitude of voltage supply is zero or close to zero and secondly from a customer perspective when there is a loss of galvanic connection between the customer and the main network. In as much as to most customers the second description does not correspond appropriately with their requirement regarding interruption, it is however the most convenient way for system operators to collage availability data for their analysis and interpretation.

The 4th benchmarking report by CEER noted that even though voltage is used in the definition of interruption, continuity data collection is based on opening and closing of interrupting devices. This means practically the earlier two descriptions of interruption are equivalent.

Power interruption could be classified in several forms, the planned and unplanned interruptions, in terms of duration as long, short or transient interruption, the component outage and incident, supply interruption as well as exceptional event. In most instances this exceptional type of event category are treated separately when analyzing continuity of electricity supply. Some generation companies do refer to this type of interruption as an Outside Management Control (OMC) outage or interruption.

Power Generation, transmission and distribution/supply companies take record of the periods of interruption and compute what is referred to as availability factor. This is simply the available periods of generating plants or electricity supply to the total period within the reporting time. This is often used for analysis, benchmarking as an aspect for their quality performance monitoring and reporting.

The table below shows the various types of interruptions that are monitored in different countries.

**Table III: Types of interruptions monitored by different Countries (data Source: CEER, 2008)**

| Country | Long interruptions | Short interruptions | Transient interruptions | Unplanned interruptions | Planned interruptions |
|---|---|---|---|---|---|
| Austria | X | | | X | X |
| Belgium (Brussels region) | X | | | X | X |
| Belgium (Flemish region) | X | X | | X | X |
| Belgium (Walloon region) | X | | | X | X |
| Belgium (Federal) | X | X | | X | |
| Czech Republic | X | | | X | X |
| Denmark | X[4] | X[4] | | X | X |
| Estonia | X | | | X | X |
| Finland | X | X | | X | X |
| France | X | X | X[2] | X | X |
| Germany | X | | | X | X |
| Hungary | X | X | X | X | X |
| Italy | X | X | X | X | X |
| Lithuania | X | X | | X | X |
| Luxembourg | X | | | X | X |
| the Netherlands | X | | | X | X[3] |
| Norway | X | X | | X | X |
| Poland | X | X | | X | X |
| Portugal | X | X[1] | | X | X |
| Romania | X | | | X | X |
| Slovenia | X | | | X | X |
| Spain | X | X | | X | X |
| Sweden | X | | | X | X |
| United Kingdom | X | X | | X | X |

(1) In Portugal, all interruptions (including short ones), are monitored at transmission level. But in accordance with the quality of service code, only long interruptions are reported.
(2) In France, the TSO monitors transient interruptions, but does not calculate any specific indicators for transient interruptions.
(3) In the Netherlands, planned interruptions are only monitored from 2006.
(4) In Denmark, all interruptions lasting 1 minute or more are monitored.

In Ghana, the generation plants measure availability of their plants. The table and figure below shows details for two generating plants in Ghana.

Table IV: Technical performance analysis of Akosombo and Kpong hydro generating stations for 2001 (Source: PURC, 2001)

| Performance Measure | 1st Q | 2nd Q | 3rd Q | 4th Q | Average / Total | PURC Benchmark | Variance |
|---|---|---|---|---|---|---|---|
| Generation Availability Factors | | | | | | | + / (-) |
| Akosombo GS | 99.2 | 98.8 | 98.8 | 95.5 | 98.0 | 95 | +3.0 |
| Kpong GS | 98.9 | 98.7 | 99.3 | 98.8 | 98.80 | 95 | +4.0 |

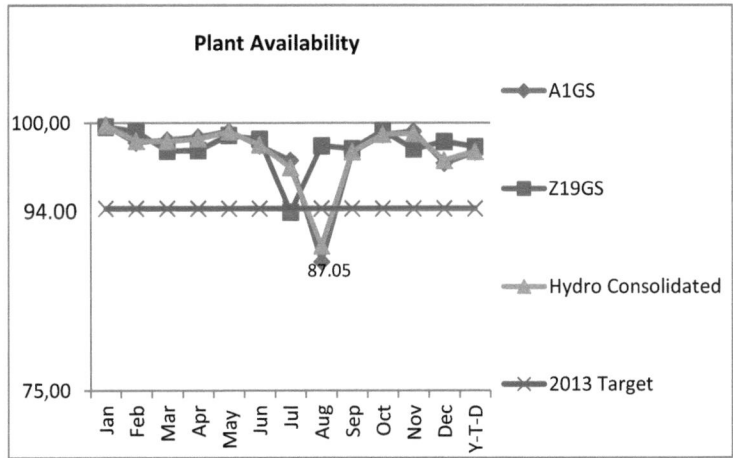

Figure 12: 2013 Annual availability factor for Akosombo and Kpong Generating Station (Source: VRA, 2014)

Legends:    AIGS - Akosombo Generating Station
            Z19GS – Kpong Generating Station

Again in Ghana, the public utility regulatory commission (PURC) sees availability as a quality performance indicator and for distribution companies it's similar to what is referred to as the network security. By network security, it analyzes the number of interruptions within every 100km of system length.

**Table V: Number of NedCo supply interruption per 100km of system length for 2001 (Source: PURC, 2001)**

| REGION | 1st Q | 2nd Q | 3rd Q | 4th Q | TOTAL NO. OF OUTAGES PER 100KM OF SYSTEM LENGTH | PURC BENCHMARK | VARIAN + / (-) |
|---|---|---|---|---|---|---|---|
| Brong – Ahafo | 3.9 | 5.8 | 3.9 | 3.1 | 16.7 | N/A | |
| Northern | 2.0 | 7.9 | 6.3 | 11.1 | 27.4 | N/A | |
| Upper East | 0.1 | 0.2 | 0.1 | 0.1 | 0.5 | N/A | |
| Upper West | 0.0 | 0.1 | 0.0 | 2.8 | 2.9 | N/A | |
| Average | 1.5 | 3.5 | 2.6 | 4.3 | 11.9 | N/A | |

**Table VI: Duration of supply hours lost per connected SLT customers for Electricity Company of Ghana in 2001 (Source: PURC, 2001)**

| REGION | DISTRIBUTION OUTAGES (HOURS) | DURATION OF OUTAGES BY VRA (HOURS) |
|---|---|---|
| | SLT | SLT |
| Accra East | 3.8 | 2.7 |
| Accra West | 0.1 | 2.3 |
| Tema | 23.0 | 0.0 |
| Eastern | 4.8 | 0.1 |
| Volta | 0.2 | 0.0 |
| Western | 3.9 | 0.6 |
| Central | 7.4 | 0.1 |
| Ashanti | 33.5 | 0.6 |
| **Average** | **9.6** | **0.8** |

The various categories of interruption such as planned, unplanned, long, short transient etc. are all monitored by various generation, transmission and distribution companies across the globe. Below is a figure showing results of interruptions monitored from 1999 to 2007 for some countriess

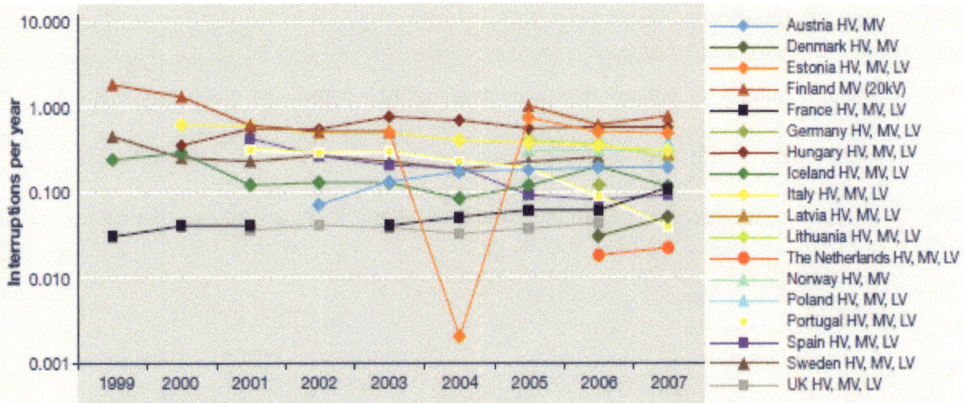

**Figure 13: Planned Interruption: Number of interruptions per year from 1999 to 2007 for various countries (Source: CEER, 2008)**

.

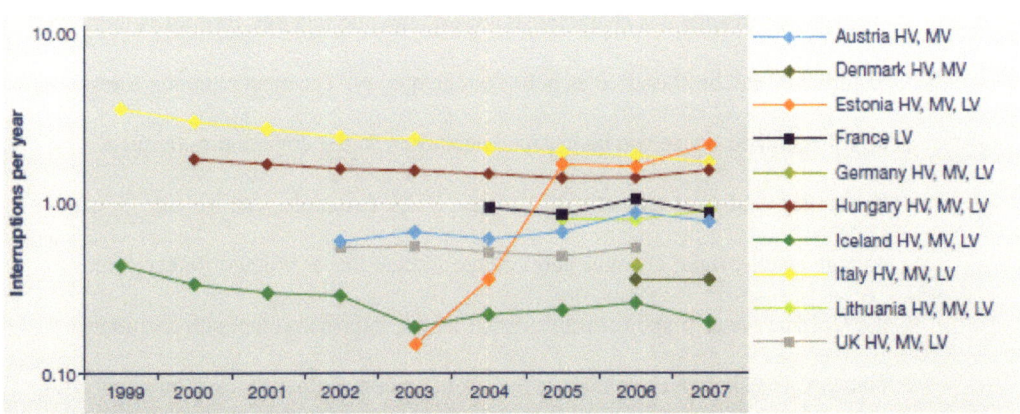

**Figure 14: Unplanned interruption per medium voltage level: Number of interruptions per year from 1999 to 2007 for various countries (Source: CEER, 2008)**

Poor quality electricity supply is therefore characterized by interruption in supply but not limited to that.

## Voltage Quality

Voltage quality is yet another aspect of quality supply of electricity. This aspect deals with the usefulness of electricity when it is available. By usefulness, the technical characteristic such as the level of voltage quality is considered.

The International Electrotechnical Commission (IEC) considered electromagnetic disturbance as any electromagnetic phenomenon that when present in an electromagnetic environment can cause electrical equipment to deviate from its intended performance.

Voltage disturbance is a characteristic of voltage and that any form of voltage disturbance can be classified as poor voltage quality. The most common forms of voltage disturbance can be grouped according to the deviation in frequency (frequency and time deviation), Voltage RMS (examples are voltage dips, voltage swells, rapid changes and voltage fluctuation or voltage flickers) and the Voltage wave shape (examples such as the inter-harmonic, sub-harmonic voltages, transient over voltage or the main signaling superimposed on supply voltage).

The 4th benchmarking report of CEER noted that European Committee for Electrotechnical Standardization and EN 50160: 2007 Voltage characteristic of electricity supplied by distribution networks serves as the norm regarding the voltage characteristic for the supply of electricity in Europe.

Voltage quality recorders are sometimes provided and installed by the distributor companies for customers to check this quality element of electricity.

The table below shows continues monitoring of voltage distribution by some European countries which also goes to attest to the fact that voltage quality is a vital element of quality electricity supply.

**Table VII: Continuous monitoring of voltage distribution by various European countries (Source: CEER, 2008)**

| Voltage disturbance | Belgium | Czech Republic | France | Greece | Hungary | Italy | the Netherlands | Norway | Portugal |
|---|---|---|---|---|---|---|---|---|---|
| Power frequency (1) | HV | HV | EHV, HV | LV | | EHV, HV | | | All |
| Supply voltage variations | HV, MV | HV | EHV, HV, MV | LV | LV | EHV, HV, MV | All | | All |
| Single rapid voltage changes | | HV | | LV | | EHV, HV, MV | All | EHV, HV, MV | |
| Flicker | HV, MV | HV | EHV, HV | LV | | EHV, HV, MV | All | | All |
| Voltage unbalance | HV | HV | EHV, HV | LV | LV | EHV, HV, MV | All | | All |
| Harmonic voltages | HV, MV | HV | EHV, HV | LV | LV | EHV, HV, MV | All | | All |
| Voltage dips | HV | HV | EHV, HV, MV | LV | LV | EHV, HV, MV | | EHV, HV, MV | All |
| Voltage swells | HV | HV | MV | LV | LV | EHV, HV, MV | | EHV, HV, MV | |
| Transient overvoltages | | HV | | LV | | | | | |
| Interharmonic voltages | | HV | | LV | | | | | |
| Mains signalling voltages | | HV | | LV | | | | | |

(1)   In all countries, the power frequency is monitored and managed by the interconnected European transmission system operators and international system operation agreements. This table only refers to what is monitored by voltage quality instruments in place for continuous monitoring.

## Commercial Quality

Commercial quality of electricity supply deals with the quality of customer service provided to the customer with regards to electricity. It includes directly associated transactions between the electricity supplier and the end user customer. These services do not end at just the supply and sale of electricity but also includes other forms of contractual agreement between the customer and the electricity company. These could include services such as verification of meters, terminating electricity supply or starting new connection altogether.

For most energy markets that are not fully liberalized, policies and implementation of this quality element of electricity supply is not effective. This notwithstanding, it still is a vital quality element of electricity supply.

For countries with fully developed competitive markets, prompt and professional handlings of complaints in relation to commercial quality are much more effective. This is so because customers whose complains are not successfully resolved have options to switch to other service providers and the regulatory measure makes service providers oblige to provide more effective remedy procedures in solving electricity quality issues.

**Table VIII: Response time to customer complaints in written form. (Source: CEER, 2008)**

| | Country | Type of standard OS or GS | Standard Quantity | Unit | Actuals in 2007 Quantity | Unit | Compensation in case of non-performance Type | Sum in EUR | Payment method | Remark |
|---|---|---|---|---|---|---|---|---|---|---|
| sp | Czech Republic | GS | 15/30 | day | | | Compensation in case of non-performance | 20 each day over limit, max. 800 | upon request | |
| | Hungary | GS | 15 | day | | | Compensation | 20 | upon claim | |
| | Italy | OS | 20 | day | 15.96 | day | | | | 90% LV 95% MV within 20 working days |
| | Latvia | OAR | 15 | day | | | | | | |
| | Lithuania | OAR | 30 | day | 15 | day | | | | |
| | Romania | OS | 30 | day | | | | | | |
| usp | Czech Republic | GS | 15/30 | day | | | Compensation in case of non-performance | 20 each day over limit, max. 800 | upon request | |
| | Estonia | OS | 15/30 | day | 15 | day | | 0 | | business 30, residental 15 |
| | Hungary | GS | 15 | day | | | Compensation | 20 | upon claim | |
| | Italy | OS | 20 | day | 15.96 | day | | | | 90% LV 95% MV within 20 working days |
| | Latvia | OAR | 15 | day | | | | | | |
| | Lithuania | OAR | | | | | | | | |
| | Portugal | GS | 15 WD | | | | Compensation | 18/30/92 * | Automatic in the bill | WD- working day; total annual compensations - € 192 |
| | Romania | OS | 30 | day | | | | | | |

In a country like Ghana where the energy market is not practically fully liberalized yet, the commercial quality of electricity supply seems not to be considered at all. A survey conducted on ten (10) categories of industries on the various element of quality of electricity supply revealed that the commercial quality of electricity supply to industries was not satisfactory. Majority of the executives and senior officers who responded to the survey questions from the various industries do not know the commercial quality aspect of quality electricity supply. About 72% of the respondent said they were unhappy with the customer service provided by the distribution company.

# PART III

## INDUSTRIAL PERFORMANCE

### Performance in Context

Performance is a widely discussed subject. Several theories have been proposed as to measuring performance right from individuals at the shop floor through the Chief Executive Officer (CEO) of an organization to the organization or industry as a whole.

Performance can be said to be a measure of how a mechanism or process achieves its objective or purpose well.

### Industrial / Organizational Performance

Organizational performance is an open question with only few studies using consistent description and measures. However, organizational performance can be said to be the degree to which organizations attain a given set of pre-defined goals or targets that are distinctive to its mission. This in a broader sense will be how well an organization is managed as well as the values that this organization delivers to customers and other shareholders. It can also be said to be the measure of output as against the organization's goals and objectives.

Organizational performance is an important criterion in evaluating organizations, their actions and settings (environments).

## Industrial / Organizational Performance Indicators

With the current increasingly competitive market, organizations competiveness becomes vital if there is any hope of survival in such as a globally competitive business environment. Various theories and principles have been populated that aims at helping organizations develop indices so as to measure the extent to which the organization is performing and if the corporate strategies set out are effective in achieving the desired output.

Performance indicators for industries are basically quantifiable measurements that reflect the company's goals and has a potential of showing the extent to which performance has been achieved taking into consideration all needed factors. These performance indicators therefore form the domain of excellence for any organization.

It is therefore important to identify the performance indicators of an organization and how it can be measured. Identifying these key performance indicators will ensure a much focused and business oriented effort.

Because key performance indicators are quantifiable measurements that mirror the critical success factors of organizations, these indicators must be well linked to the organization's objectives as they are used as yard sticks to enable shareholders determine if the organization is on the desired path and how well it is performing in relation to their strategic goals and objectives.

Measurement and evaluation of the key performance indicators of an organization are done by means of data. It is therefore equally important that these data are analyzed well so as to comprehend the complete or overall performance of the organization.

Organizational performance indicators vary greatly from one organization to another. This is largely due to indicators being linked to the goals of the organization. The environment where the industry is located, type and nature of the industry are all some other factors that frames performance outcome hence making organizational performance indicators vary greatly.

However four (4) specific areas that are generally covered are:

1. **Financial Performance:** Both small and large firms have indicators that measure how financially sound a business is. Even non-profit organizations measure to see if they break even. Therefore organizational outcomes such as profits, return on investment and return on asset are some of the indicators used to determine the financial performance aspect of an organization.

2. **Product / Market Performance:** Organizations set indicators to measure their internal businesses processes in relation to their products and/or market performances. Sales, market shares, quality standard they have achieved, sigma level, number of products produced are

some of the indicators organizations measure to determinate their product/market performance in relation to the organizational performance.

3. **Stakeholders satisfaction:** Satisfaction of the stakeholder is yet another key area that performance indicators measure. Indicators such as customer satisfaction, employee satisfaction, customer return rate are used to determine the stakeholders' satisfactory aspect of an organization in relation to the organizational performance.

4. **Shareholders return:** Performance of an organization's stocks, returns on stocks and shares, economic value added, total shareholder returns are some indicators to measure shareholder return and organizational performance as a whole.

**Measuring Organizational Performance**

Publications on performance measurement dates as far back as in the 1980s. In the twenty first century, researches on measuring organizational performance were examined more generally.

Measuring and analyzing organizational performance have several importance and key among them is its ability to turn organizational goals into reality thereby setting the organization on the desired path.

Measuring performance of an organization will require weighing the relevance of such performance to the key stakeholders. The nature of measuring performance of an organization can be firm specific. This could depend on the internal policies of that firm and management strategy.

Considering the fast changing business environment that we have now and how organizations compete among themselves, organizational performance varies at different rates and times for industry, corporate and businesses. It is therefore important that organizational performance measures are time specific. Therefore the time specific factor must also be considered when measuring performance of an organization.

Organizations in general both small and larger ones use both financial and non-financial measures in determining their performance as an organization. Generally, smaller firms however place more weight on product performance and profitability. In the United Kingdom, smaller firms give less emphasis on their overall profitability but rather weight highly on their debt levels.

Larger organizations most often than not put more weight on the financial measures than the non-financial ones.

# PART IV

# IMPACTS OF ELECTRICITY ON INDUSTRIAL PERFORMANCE

Electricity is essential to every modern society. Electricity has serious impacts not just directly on industries but the entire economy of any country whether developed or developing country. Most developing countries however turn to be more affected by poor quality electricity supply than developed countries. Nonetheless the positive impact that electricity makes to economic development of any country cannot be overemphasized. Electricity is used directly in running industrial machines and it contributes greatly to productivity of human capital.

**The Case of Africa**

In 2011, power outages in Senegal has impacted negatively on business productivity. A research data on 528 businesses in Senegal showed that 57% of businesses had electricity as a major constraint to their productivity. It has been documented that a typical month in Senegal registered almost 26 outages on average, with an outage lasting over 2 hours. These outages result in an average of 5.1% loss of annual sales in Senegal.

The case of Nigeria is even worse. Power outages in Nigeria in a typical month averages a little over 26 times and a typical duration of outage is 8.2 hours. The lost due to an outage in Nigeria is 8.9% of annual sales. All these figures are above averages for both Sub-Saharan Africa and the world. About 75.9% of firms in Nigeria identify electricity as a major constraint to their businesses. It is evident that poor quality electricity supply in Nigeria slowed industrial growth.

The table below shows details of the electricity outage in Senegal, Nigeria, Sub-Saharan Africa and the World Average.

**Table IX: Electricity Outages (Source: Cissokho et al, 2013)**

| Indicator | Senegal* (2011) | Nigeria | Sub-Saharan Africa | World Average |
|---|---|---|---|---|
| Number of electrical outages in a typical month | 25.8 | 26.3 | 10.7 | 8.6 |
| Duration of a typical electrical outage (hours) | 2.3 | 8.2 | 6.6 | 4.0 |
| Losses due to electrical outages (% of annual sales) | 5.1 | 8.9 | 6.7 | 4.8 |
| Percentage of firms owning or sharing a generator | 90.7 | 85.7 | 43.6 | 31.6 |
| Proportion f electricity from a generator (%) | 30.8 | 47.5 | 13.88 | 7.1 |
| Percentage of firms with electricity as a major constraints | 57.5 | 75.9 | 50.3 | 39.2 |

For Ghana, just like many other countries, energy is one of the fundamental catalysts needed for its rapid industrialization and subsequent development. However, poor quality of electricity supply costs the country so much. Within the third quarter of 2012, the combined effects of frequent power outages and regular power surges led to the rise of unit cost of production. Poor electricity supply was ranked the topmost overall challenge in Ghana from July 2012 to September 2012.

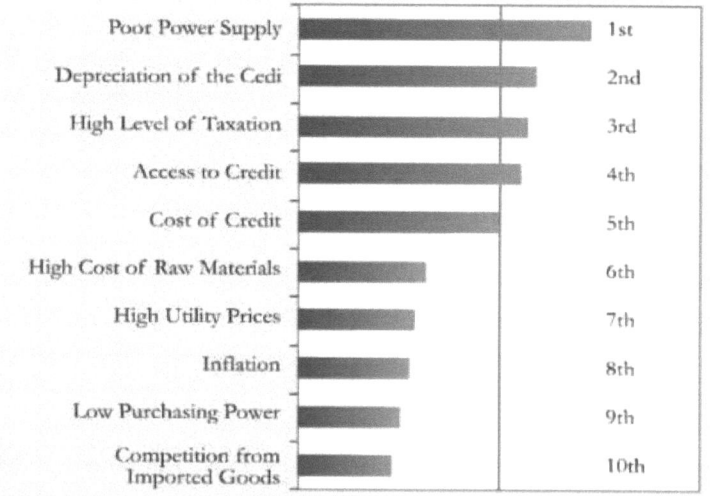

**Figure 15: 2012 3rd Quarter Overall Challenges in Ghana (Source: GB & F Magazine, 2012)**

A report by the Centre for Policy Analysis (CEPA) in 2007 stated that Ghana was estimated to have lost in excess of 14 million Ghana cedis in revenue as result of decline in production in various sectors including manufacturing, mining and quarrying. This decline as the report noted was due to the rising production costs and loss of productive man-hours on account of the energy crisis.

Guinness Ghana noted that 4% decline in volumes of Ghana's only total beverage business and a Diageo business for the first half of 2013 was due to inconsistent supply of electricity.

Power outages cost manufacturing companies in Africa about 5 to 6 % of their revenue.

**The Case of America**

Electricity is known to have a unique ability of conveying both energy and information thereby yielding increases in products, factor's applications etc. United State of America department of energy in 2003 noted how electricity had positively impacted the United State economy. This according to the report noted how electricity aided in economic growth of the United State. The figure below shows how electricity impacted positively on the United States' economy.

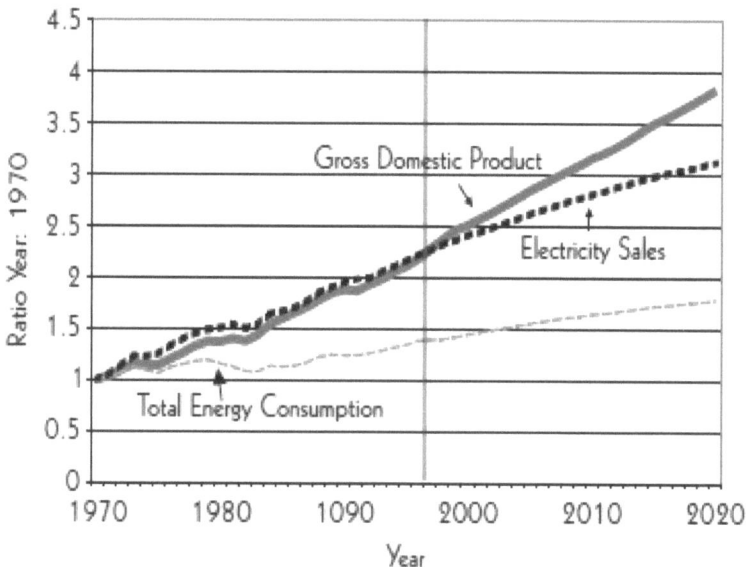

Figure 16: Electricity and Economic Growth in U.S. (Source: U.S. Department of
Energy Transmission Reliability Multi-year Program Plan, 2013)

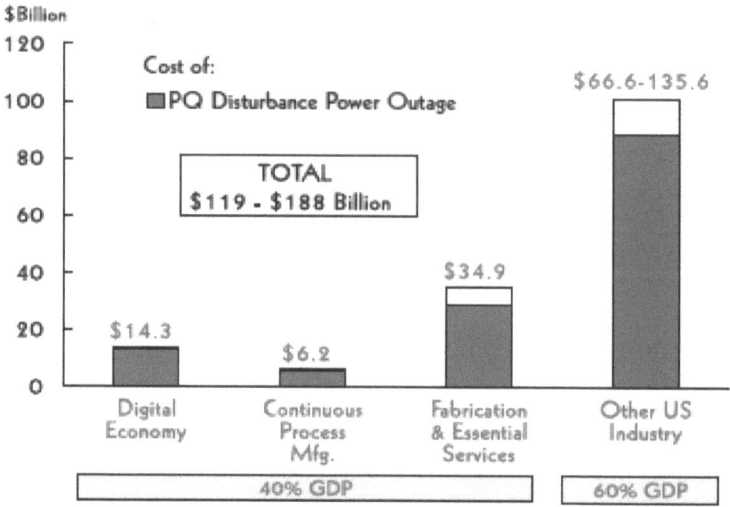

Figure 17: Annual Cost of Electricity Power Outage and Power Quality Disturbances
(Source: U. S. Department of Energy, 2003)

Power outages and power quality disturbance on the other hand results in an estimated cost $ 25 to $180 billion annually to the economy. Figure 17 above shows the annual cost of electricity outage and power quality disturbances.

Duke power survey's (1992) data on the cost of power outage to the industrial sector in the southeastern region of the United States showed that an annual estimated cost of outage on the industrial sector alone was 59 billion dollars.

Another data by Primen Inc. (2001) from 985 industrial and Digital Economy (DE) firms also estimated the cost of power outages for all the geographical regions of United States. The estimated cost for just power outage for industrial and digital economic firm was $132 billion to $ 209 billion.

Table 9 below shows previous estimated Annual cost of Power outages.

**Table X: Previous estimate of annual cost of power outage (Source: Executive Office of the President, 2013)**

Previous Estimates of Annual Cost of Power Outages

| Source | Estimate (2012 dollars) | Year Published |
|---|---|---|
| **All Outages** | | |
| Swaminathan and Sen | $ 59 billion | 1998 |
| PRIMEN | $132 to $209 billion | 2001 |
| LaCommare & Eto | $28 to $169 billion | 2005 |
| **Weather-related outages** | | |
| Campbell (CRS) | $25 to $70 billion | 2012 |

In 1992, power rationing in Columbia was estimated to reduce the overall country's economic output by nearly 1% of GDP.

**The Case of the Middle East and Asia**

The overall economic cost of power outage for 843 firms in the industrial sector of Pakistan was about 8.8% of the added value by the industrial sector in 1984 – 1985. The most affected industries are the food, beverages and tobacco, the machinery and equipment industries, the metal and metal product industries and the textile industries. Power outages in the industrial sector led to a 1.8% decline in the overall GDP in 1984 - 1985.

In India, expansion of electricity networks helped in the entry and performance of small firms. Considering the magnitude of this positive impact on expanding electricity network in India, a more serious policy consideration to expand electricity to networks including those in the rural areas are being considered so as to promote industries and economic development.

**The European Case**

On Sunday September 28$^{th}$, 2003, all the provinces in Italy except Sardinia experienced a large scale interruption in electricity supply which affected over 55 million people and lasted between 3 to 16 hours depending on the providence. The macroeconomic damage of this outage was documented to be € 1,182 million. The losses to the agricultural sector, production of goods and services was € 897.5 million whilst losses to households amounted € 285 million.

The figure below shows the outage duration in the various affected regions in Italy and table 7 provides details on losses across the regions, sectors and households.

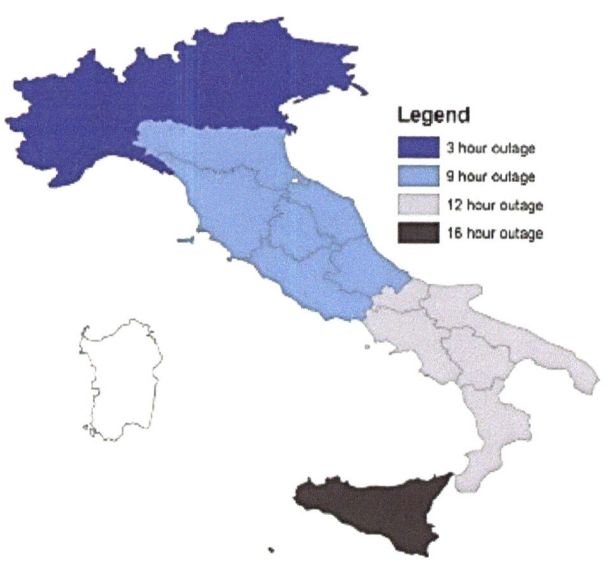

Figure 18: September 28th 2003 Power Outage in Italy lasting 3 - 16 hours
(Source: CIRED, 2014)

**Table XI: Total loss across all regions, sectors and households relevant to outage durations in million euros (Source: CIRED, 2014)**

| Economic Sector | Agriculture (Primary Sector) | Production of Goods | Services | Households | Total Losses Inc. WTP Household |
|---|---|---|---|---|---|
| North | 5.3 | 136.7 | 60.8 | 43.1 | 245.8 |
| Center | 20.6 | 217.6 | 154.6 | 98.2 | 491.0 |
| South | 20.9 | 82.8 | 97.6 | 94.3 | 295.5 |
| Sicily | 12.4 | 33.7 | 54.6 | 49.5 | 150.1 |
| Total | **59.2** | **470.8** | **367.5** | **285.1** | **1182.4** |
| % of GDP | 0.004% | 0.03% | 0.03% | 0.02% | 0.083% |

Table XII: Summary estimates of the cost of power outages (Source: Musilium O. Oseni)

| tudy | Scope | Method / Data | Focus | Findings/Outage cost estimates in 2007 prices |
|---|---|---|---|---|
| ental & avid 982) | US & Israel | Data on firms' average outage duration in 1980. -US: 10 hours p.a. -Israel: 70 hours p.a. Marginal cost approach. | Computation of marginal outage costs. | Reliability varies directly with outage costs. Outage cost: Israel - US$0.40/kWh |
| asha et . (1989) | Pakistan | Nationwide random survey of 843 firms in 1984/85 Reported loss data - Planned outages - Unplanned outages | Computation of output loss due to power outages, Computation of the multiplier effects of firms' loss due to outages on the overall economy. | Overall outage cost accounts for 8.8% of industrial output value added. Off which - Planned outages: 65% - Unplanned outages: 35% Outage multiplier: 1.34 Overall impact on GDP: 1.8% reduction Outage costs per kWh: - Planned: $0.58 - Unplanned: $1.02 |
| eenstock 991) | | Self-assumed data Risk- averse model | Calculation of outage cost under risk-averse behaviour. | Reliability varies indirectly with outage cost. The higher the probability of outage, the greater the investment in backup (i.e., investment in generator is sensitive to outage risk). |
| aves et . (1992) | US | Use of data on interruptible service schemes. - 8 participants - 11 non-participants | Estimation of shortage cost. | None of the parameter estimates was significant. Expected outages costs decrease as the size of the interruption increases. Shortage cost (Utility): $4.63 - $5.58/kWh. Outage cost for Industry: $6.97-$34.85/kWh. |
| latsukaw & Fujii 994) | Japan | 1988 Survey of backup among industrial & commercial consumers with large computers by CRIEPI, Japan Mailed questionnaires - Sample: 2,200 -Complete questionnaire Returned: 236 Discreet choice model | Computation of outage costs using back-up data, Evaluation of the factors affecting the demand for back-up. | Demand for backup varies inversely with reliability & user costs of backup investments Customers face trade-off between price & reliability of power supply Customers characteristics have significant effects on backup investment Outage cost: $50.72 - $236.17/KW |
| eenstock : al. 997) | Israel | Surveying of 794 business and public sectors. -Data on backup -Reported losses -Firms characteristics Two-limit tobit model. | Separation of total outage cost from unmitigated cost, Comparison between the computed costs from revealed datasets and subjective datasets analyses | Outage cost - $9.21/kWh Unmitigated cost- $3.45/kWh Total annual cost – $45.34/KW Back-up rate – 33 percent Reliability varies inversely with demand for backup and total outage cost, but varies directly with marginal cost per kWh unsupplied. |
| teinbuks : Foster 2010) | Africa | Use of firms' datasets on back-up & sale losses - 25 countries - 8483 firms - dataset between 2002 & 2006 Probit & tobit models Marginal cost method | Computation of outage costs, Investigation of drivers of auto-generation, Evaluation of cost-benefits with focus only on sale loss reduction. | Impact of power unreliability on demand for generator is limited Outage cost varies directly as reliability The cost-benefit of self-generation is not significant Outage cost: $0.13 - $0.76/kWh |

47

## The Six (6) General Areas that Poor Quality Electricity Impacts

In general, most of the impacts electricity has on industrial performance are valued in monetary term. The negative impacts vary for both planned and unplanned with unplanned power outage mostly more severe than those that are planned. Poor quality electricity supply is a drag on industrial performance. This may however vary from industry to industry depending on how heavenly dependent the operations/activities of a firm is on electricity.

Generally, electricity supply impacts both positively and negation on these six main areas of industries performance. These are:

1. Product or service quality
2. Finances of the industries
3. Volume/quantity of products/services produced.
4. Asset life and return on asset/investment
5. Customer satisfaction
6. Shareholder return

# REFERENCES

Adegbemi B. Onakoya, (2013) Energy consumption and Nigerian Economic Growth: An Empirical Analysis

Aditya Parida and Uday Kumar (2009). Handbook of Maintenance Management and Engineering,

Adom, P. K., Bekoe, W., and Akoena, S. K. K.(2011). Modelling aggregate domestic electricity demand in Ghana: An autoregressive distributed lag bounds cointergration approach. 2011 Elsevier Ltd

Allcott, H., Collard-Wexler, A., and O'Connel S. D. (2014). How do electricity shortages affect productivity? Evidence from India. March 10, 2014.

Anyomi S. Dennis, (2014) Assessing the impact of Electricity on Industrial Performace: A case Study of Industries in the Greater Accra Region of Ghana.

Bibhuti Bhusan Mahapatro, Humana Resource Management. Page 272 - 279

Boland Justin (2005). California Institute of Technology, Pasadena, California: Micro Electrit Power Generators. Page 5, 6, 7, 9

Boyd, B. K., Gove, S. and Hitt, M. A. (2005). Construction measurement in strategic management research: Illusion or reality? Strategic management Journal, pages 239 -257

Chakravarthy, B. S. (1986). Measuring strategic performance. Strategic management Journal. Pages 437 – 458.

Cired Workshop (2014), Assessing the Socio- Economic Effects of Power Outage in the European Union Ad Hoc. Using www.blockoutsimulator.com.

Copenhagen (2011). Organizational Key Performance Indicators – a management tool with bottom line effect. Human Capital Management.

Davig, W., Elbert, N., and Brown, S., (2004). Implementing a strategic planning model for small manufacturing companies: An adaptation of the balanced scored S.A.M Advanced Management Journal, Page 18 – 25

Dr. Karl Albrecht (2011), Organizational Performance. Meeting the Challenges of the New Business Environment.

Dutta, S., and Reichelstein, S. (2005). Stock price, earnings, and book value in managerial performance measures. The Accounting review.

Electricity Company of Ghana (2014). www.ecgonline.info (retrieved on March 7, 2014)

Energy Commission, Ghana, National Energy Statistics, July 2013,   Pg 8, 16 & 17

Energy Foundation, (2014), Energy in Ghana

ERGEG (2007), Towards Voltage Quality Regulation in Europe.

Executive Office of the President (2013). Economic Benefits of Increasing Electric Grid Resilience to Weather Outages. August 2013. Page 18

Ghali, K. H. and El-Sakka,   (2004) M.I.T. Energy Consumption and Output growth in the Canada: Multivariate Cointegration analysis. Energy Economics pages 26, 225 – 238

Ghana Business & Finance Magazine, 2012, Ghana's Power Challenges and National Productivity. December, 2012. www.ghanabizmedia.com

Ghana Grid Company Limited (2013), 2013 Electricity Supply Plan, Page 6

Ghana Web (March 4, 2013), Ghana's Energy Crisis, The CPP's Blue Print

Ghanaweb (2013), Poor water, electricity supply affect Guinness Ghana profits. Business News of Thursday, October 31, 2013. www.ghanaweb.com.

Global Competitive Enterprise, Resource Centre.   (Retrieved on March 14, 2014) https://nationalvetcontent.edu.au/alfresco/d/d/workspace/SpacesStore/6c499386-f2e0-4612-943e-317149ff1011/10_04/toolbox/resources/res4040/res4040.htm

Hamit- Hagger (Greenhouse gas emission, energy consumption and economic growth: a panel co-integrated analysis from Canadian industrial sector Perspective

Hawawini, G., Subramanian, V., and Verdin, P. (2003). Is performance driven by industry – or firm – specific factors? A new look at the evidence. Strategic management Journal, Pages 1 – 16

Iana Vassilena (2012). Characteristic of Household Energy Consumption in Sweden: Energy Savings Potential and Feedback Approaches

Imani, (January 30th, 2014) Pricing and Deregulation of the Energy Sector in Ghana: Challenges & Prospects

International Energy Agency, (2010). Renewable Energy Essentials: Hydropower. www.iea.org

Jing Li and Yu Xue (2010), The Cold-Powered Electricity Market in China, University of Gravle, Department of Energy systems, Pages 14, 19

Kessides, C. 1993. The contributions of infrastructure to Economic Nigeria in focus. Global issue papers: No 12 Henrich Boll Stiftung.

Kirby, J. (2005). Toward a theory of high performance. Harvard Business Review, July – August. Pages 30 -39

Laitinen, E. K. and Chong, G. (2006). How do small companies measure their performance? Problems and Perspectives in Management. Page 49 -68

Lee, C. C. and Chiang, C. (2008), Energy Consumption and Economic growth in Asian Countries: A more Comprehensive analysis using Panel data. Resource and Energy economics pages 30, 50 – 65

Malina, M. A. and Selto, F. H. (2004). Choice and change of measures in performance measurement models. Management Accounting Research. Pages 441 – 469

Maria Micallef, Key Performance Indicators for Business Excellence. Partner RSM Malta

McGahan, A. M. and Porter, M. E. (1997). How much does industry matter, really?: Strategic management Journal . Pages 15 - 30

Ministry of Energy and Petroleum (January 2014), Sectorial Overview

Moullin, M. (2003). "Defining Performance Measurement." Perspectives on

Musiliu O. Oseni. Power Outages and Cost of Unsupplied Electricity: Evidence from Backup Generators among Firms in Africa.

Narayan, P. K., Smyth, R. (2008). Energy consumption and real GDP in G7 countries, New evidence from panel cointegration with structural breaks. Energy Economics, pages 30, 2331-2341.

Northern Electrification Department Company (2014). www.nedco.com.gh (retrieved on March 7, 2014)

Oshodi, A. F and Oloni, E. F (2007). Public – Private Partnership: A Publishing (London: 2004)

Pasha, H. A., Ghaus, A., and Malik, S. (1989). The economic cost of power outages in the industrial sector of Pakistan, Energy Economics. Pages 301 -318.

Performance 2(2): 3.

Pina Andre, Silva Carlos, Ferrao Paulo (2011), The impact of demand side management strategies in the penetration of renewable electricity. 2011 Elsevier

Popova, V., and Sharpanskykh, A. (2009). Modeling organization performance indicators. 2009 Elsevier B. V.

Public Utilities Regulatory Commission (2001), Annual Report 2001/ Pages 26 – 30.

Richard, P. J., Devinney, T. M., Johnson, G., and Yip, G. S. (2009) Measuring Organizational Performance: Towards Methodological Best Practices. Journal of Management, 2009 Review Issue.

Rud, Juan Pablo (2011), Electricity provisoin and industrial development: Evidence from India. Department of Economic, Royal Holo way, University of London. UK. 2011 Elsevier B.V.

Rumelt, R. P. (1991). How much does industry matter? Strategic Management Journal, Pages 167 – 185

Sari R. and Soytas U. (2007). The growth of Income and Energy Consumption in Six Developing Countries. Energy Policies. Pages 35, 889-898.

Sersen, E & Vorsic, J. Quality of Electricity Supply as Service

Shahbaz M. and Lean, H. H. (2012). The dynamics of electricity consumption and economic growth: A revisit study of their causality in Pakistan

Tang W, Li Z., Qiang M., Wang S., Lu Y., (2013), Risk management of hydro power development in China, 2013 Elsevier Limited

U. S. Energy Information Administration (2009), Independent Statistic & Analysis

U. S. Energy Information Administration (2010), International Energy Outlook 2013, Independent Statistic & Analysis. Pages. Page 93, 94,95, 97, 105

U. S. Energy Information Administration (2014), Energy Kids. (Retrieved on March 8, 2014) http://www.eia.gov/kids/energy.cfm?page=electricity_home-basics-k.cfm

Unite States Department of Energy Office of Elec Transmission and Distribution (2003). Grid 2030, A National Vision for Electricity's Second 100 years. July, 2003. Page 3, 5, 6

Venkatraman, N. and Ramanujam, V. (1986), Measurement of business performance in strategy research: A comparison of approaches. Academy of Management Review. Pages 801 – 804

Vivien, F., Tjaarda S. V. L., Briceno-Garmendia, C., John, D. C., Goddard, G. Mills, R., & Smits K. (2008). Africa's Power Supply Crisis: Unraveling the Paradoxes, World Bank publication.

Volta River Authority, (2013), Chief Executives News Letters

Volta River Authority, http://vraghana.com/resources/facts.php

Wayne Ma, (January 20th, 2014), The Wall Street Journal, China. (Economy and Business).

Wolde-Rufael, Y, 2008. Energy consumption and Economic growth: The experience of African countries Revisited, Elsevier B. V.

Wolde-Rufael, Y., 2008. The long-run relationship between petroleum imports and economic growth: the case of Cyprus. Resources, Energy and Development 5, 95 –103.

Wu Donglin (2009). Measuring Performance in Small and Medium Enterprises in the Information & Communication Technology Industries. Page 8

Wustenhagen and Menichetti, (2011), Strategic Choice for renewable energy investment: Conceptual framework and opportunity for further research. 2011 Elsevier Ltd.